# 恐龙斗士
## Dinosaur Warriors

[英] 露丝·欧文/著

刘颖/译

汉英对照
恐龙科普

江苏凤凰美术出版社

# 全家阅读
# 小贴士

★ 每天空出大约10分钟来阅读。

★ 找个安静的地方坐下，集中注意力。关掉电视、音乐和手机。

★ 鼓励孩子们自己拿书和翻页。

★ 开始阅读前，先一起看看书里的图画，说说你们看到了什么。

★ 如果遇到不认识的单词，先问问孩子们首字母如何发音，再带着他们读完整句话。

★ 很多时候，通过首字母发音并听完整句话，孩子们就能猜出单词的意思。书里的图画也能起到提示的作用。

## 最重要的是，感受一起阅读的乐趣吧！

扫码听本书英文

# Tips for Reading Together

• Set aside about 10 minutes each day for reading.

• Find a quiet place to sit with no distractions. Turn off the TV, music and screens.

• Encourage the child to hold the book and turn the pages.

• Before reading begins, look at the pictures together and talk about what you see.

• If the child gets stuck on a word, ask them what sound the first letter makes. Then, you read to the end of the sentence.

• Often by knowing the first sound and hearing the rest of the sentence, the child will be able to figure out the unknown word. Looking at the pictures can help, too.

## Above all enjoy the time together and make reading fun!

# Contents 目录

# 来战斗吧!
## Time to Fight!

6600万年前，一群三角龙正在吃草。
它们的首领是一头高大的雄性三角龙。
这时，首领发现了另一头雄性三角龙。
这头三角龙也想成为首领。

66 million years ago a **herd** of Triceratops was
eating grass.
The leader of the herd was a big male.
Then he saw another male Triceratops.
The new male wanted to be the leader of the herd.

三角龙
**Triceratops**
**(try-SERA-tops)**

5

# 用角战斗
## Horns for Fighting

这两头三角龙用角撞击对方，战斗开始了。

The two Triceratops rammed their horns together and began to fight.

战斗的胜利者将成为首领。

The winner of the fight would become herd leader.

三角龙英文名的字面意思
是"长着3个角的脑袋"。
The name Triceratops means
"three-horned head".

# 三角龙的战斗
# Triceratops Fights Triceratops

科学家发现了三角龙的头骨化石。
头骨上有洞。
这些洞是被另一头三角龙的角戳
出来的！

**Scientists** have found **fossils** of
Triceratops skulls.
The skulls have holes in them.
The holes were made by the horns of
another Triceratops!

三角龙的龙角化石
**fossil Triceratops horn**

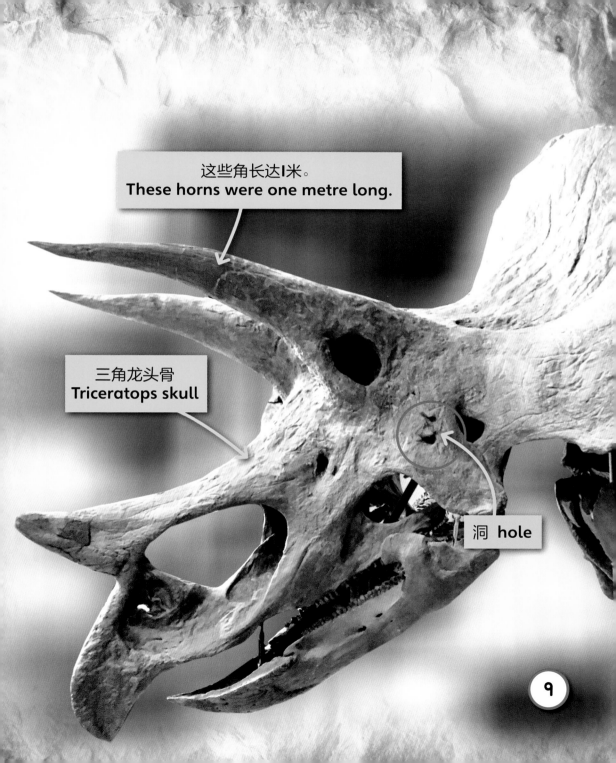

这些角长达1米。
**These horns were one metre long.**

三角龙头骨
**Triceratops skull**

洞 **hole**

9

# 剑角龙的战斗
## A Stegoceras Fight

剑角龙是一种食草性恐龙。

它和狗一样大。

Stegoceras was a plant-eating dinosaur.

It was the same size as a dog.

它的头部呈半球形，

像自行车头盔。

It had a **domed** head,

like a bike helmet.

剑角龙头盖骨
**Stegoceras skull**

雄性剑角龙在战斗时会用头撞击其他雄性。

The male Stegoceras used its head to ram other males.

科学家认为剑角龙为争夺雌性而战斗。
Scientists think Stegoceras fought over females.

剑角龙
**Stegoceras**
(steg-uh-SAIR-us)

11

# 用刺战斗
# Fighting with Spikes

剑龙用它带刺的尾巴来击退其他恐龙。
科学家发现了一块异特龙化石，化石上有一个洞，
与剑龙尾巴上的尖刺相吻合。

The Stegosaurus used its spiky tail to fight off other dinosaurs.
Scientists found an Allosaurus fossil with a hole in it that matched
the spike on a Stegosaurus tail.

异特龙
Allosaurus
(AL-oh-SAW-rus)

剑龙的尾刺长达90厘米。
The tail spikes were 90 cm long.

尾刺 spike

剑龙
Stegosaurus
(STEG-oh-SAW-rus)

13

# 咬一大口
## A Big Bite

数千万年前，
一头霸王龙攻击了一头埃德蒙
顿龙。
它从埃德蒙顿龙的尾巴上咬下
了一大块肉。

Tens of millions of years ago,
a Tyrannosaurus rex attacked
an Edmontosaurus.
It took a bite out of the tail of the
Edmontosaurus.

霸王龙牙齿 T. rex tooth

霸王龙的牙齿呈锯齿形，像锯子
一样。
T. rex's teeth were **jagged** like a saw.

霸王龙
**Tyrannosaurus rex**
**(tie-RAN-oh-SAW-rus rex)**

埃德蒙顿龙
**Edmontosaurus**
**(ed-MONT-oh-SAW-rus)**

15

# 霸王龙的战斗
## T. Rex Fights

科学家发现了一块埃德蒙顿龙化石。
化石里有一颗折断的霸王龙牙齿。

Scientists found a fossil from an Edmontosaurus.

There is a broken-off T. rex tooth in the fossil.

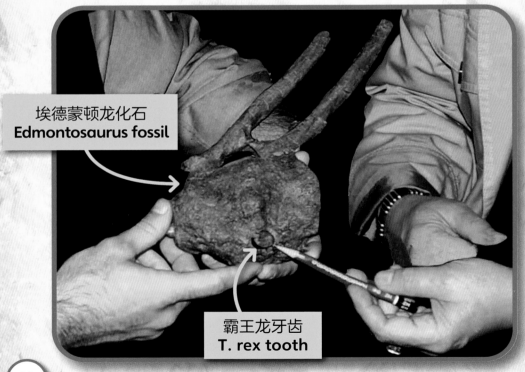

埃德蒙顿龙化石
**Edmontosaurus fossil**

霸王龙牙齿
**T. rex tooth**

霸王龙相互战斗。

一些霸王龙头骨上还留下了其他霸王龙的齿痕。

T. rex fought each other.

Some T. rex skulls have T. rex tooth marks in them.

霸王龙头骨 **T. rex skull**

齿痕 **tooth marks**

# 为食物而战
## Fighting for Food

恐爪龙会吃其他恐龙的尸体。

Deinonychus ate the dead bodies of other dinosaurs.

恐爪龙
**Deinonychus**
**(dine-ON-ee-kuss)**

体形大的恐爪龙为了获取食物也会杀死年幼的恐爪龙。

The big Deinonychus would also kill the little Deinonychus to get
to the food.

科莫多巨蜥
**Komodo dragon**

今天，科莫多巨蜥也以同样的方式争夺食物。
Today, Komodo dragons fight over food in the same way.

# 永远的战斗
# Fighting Forever

一天，一头伶盗龙用爪子刺进了一头原角龙的脖子。

原角龙咬住了伶盗龙的上肢。

两头恐龙都死了。

One day a Velociraptor stabbed its claw into the neck of a Protoceratops.

The Protoceratops bit the Velociraptor's arm.

Both dinosaurs died.

8000万年后，科学家发现了它们扭打在一起的化石。

80 million years later, scientists found their fossils together.

原角龙
Protoceratops
(PRO-toe-SERA-tops)

伶盗龙
**Velociraptor**
**(vel-OS-ee-rap-tor)**

# 词汇表 Glossary

**半球形　domed**

形似球的上半部分。
Shaped like the top
half of a ball.

**化石　fossil**

存留在岩石中几百万年前
的动物和植物的遗体。

The rocky remains of an
animal or plant that lived
millions of years ago.

## 兽群　herd

生活在一起的一大群动物。

A large group of animals that live together.

## 锯齿形　jagged

由尖角组成的不平整边缘。

With a rough edge made up of sharp points.

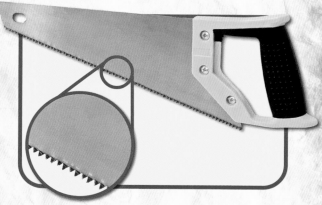

## 科学家　scientist

研究自然和世界的人。

A person who studies nature and the world.

# 恐龙小测验 Dinosaur Quiz

① 为什么有些恐龙会战斗？
Why did some dinosaurs fight?

② 我们是如何知道三角龙
相互攻击的？
How do we know that Triceratops
fought each other?

③ 剑角龙用它的半球形
头部做什么？
How did Stegoceras use its
domed head?

④ 霸王龙之间会发生战斗吗？
Did some T. rex fight other T. rex?

⑤ 如果三角龙和霸王龙打起来，你认为谁会赢？
Who do you think would win a fight between Triceratops
and T. rex?

24